FLORA OF TROPICAL EAST AFRICA

PRIMULACEAE

P. Taylor

Annual or perennial herbs, or rarely shrubs ; stems erect or prostrate and rooting at the nodes. Leaves basal or cauline, alternate, opposite or verticillate, simple or lobed, entire or dentate ; stipules absent. Flowers axillary and solitary, racemose, verticillate or paniculate, hermaphrodite, regular, sometimes heterostylous. Calyx gamosepalous, free or rarely adnate to the ovary, 4–9-partite, usually persistent. Corolla gamopetalous, rotate, hypocrateriform or campanulate, with a very short to long tube and 4–9-partite limb, or corolla rarely absent. Stamens equal in number to and inserted opposite the corolla-lobes. Ovary superior or rarely half inferior, 1-locular ; ovules 2–3 or usually many on a free central placenta. Fruit a capsule, valvate or circumscissile or rarely indehiscent. Seeds small, 1, few or usually numerous, often angular.

A largely temperate family with a few genera extending into the tropics (mainly in the mountains and highlands). The genus *Primula* L. is represented in tropical Africa by a single species which occurs in Ethiopia and Somaliland.

Ovary superior ; staminodes absent :
  Plant pubescent ; leaves lobate-dentate ; anthers apiculate ; capsule valvate or indehiscent . . 1. **Ardisiandra**
  Plant glabrous ; leaves entire ; anthers obtuse or acute, not apiculate :
    Capsule valvate :
      Robust perennials ; corolla (in tropical African species) pink or white ; flowers in terminal racemes ; leaves alternate or subopposite . 2. **Lysimachia**
      Small annual ; corolla yellow ; flowers solitary, axillary ; leaves opposite . . . . 3. **Asterolinon**
    Capsule circumscissile or indehiscent ; corolla pink, white, red or blue ; flowers solitary, axillary, or in terminal racemes . . . . . 4. **Anagallis**
Ovary half inferior ; staminodes present ; pedicels geniculate . . . . 5. **Samolus**

1. ARDISIANDRA

Hook. f. in J.L.S. 7 : 205, t. 1 (1864) ; P. Tayl. in K.B. 1958 : 146 (1958)

Pubescent herbs, creeping and rooting at the nodes or more or less acaulescent. Leaves petiolate, alternate, ovate to orbicular, coarsely lobed and toothed. Flowers small, pedicellate, bracteate, in few-flowered short extraaxillary sessile or pedunculate racemes ; bracts linear, situated at or near the base of the pedicels. Calyx-segments 5 (rarely and abnormally up to 7)

± ovate, acute, ciliate, sometimes somewhat accrescent. Corolla campanulate ; lobes 5, about ⅓ of the length of the corolla, sometimes ciliate. Stamens shortly connate at the base into a ring which is adnate to the corolla ; filaments short ; anthers apiculate. Ovary globose, setulose. Capsule globose, dehiscing by 5 short cartilaginous valves, or indehiscent. Seeds 3-angled, minutely papillose.

The genus is confined to the mountains of tropical Africa. It is very closely related to *Cortusa* L. with a few species in central Europe and Asia.

Fruit dehiscing by 5 valves :
  Corolla 2–3 times as long as the calyx, bright pink ;
    peduncle up to 8 cm. long .   .   .   . 1. *A. primuloïdes*
  Corolla scarcely longer than the calyx, white ;
    peduncle less than 1 cm. long .   .   . 2. *A. sibthorpioïdes*
Fruit indehiscent .   .   .   .   .   . 3. *A. wettsteinii*

1. **A. primuloïdes** *Knuth* in E.J. 53 : 316 (1915) ; Weim. in Svensk Bot. Tidskr. 30 : 36 (1936) ; P. Tayl. in K.B. 1958 : 146 (1958). Type : Tanganyika, Rungwe District, Kibila Gorge, *Stolz* 924 (UPS, iso.!, K, iso.!)

Stem slender, 1–1·5 mm. thick, up to 10 cm. long (rarely up to 20 cm.), with numerous persistent leaf-bases. Leaves 3–7 ; petiole slender, 12–50 mm. long, densely pubescent ; lamina ovate or oblong-ovate, up to 45 × 30 mm., sparsely pubescent on both surfaces, acute, basally cordate, coarsely double-dentate. Racemes long-pedunculate, 2–10-flowered ; axes up to 10 mm. ; peduncles up to 8 cm. long, pubescent ; pedicels filiform, pubescent, up to 18 mm. long. Bracts about 3 mm. long, lanceolate. Calyx-segments ovate, hastate or cordate, acute, ciliate, about 2 mm. long at anthesis, somewhat accrescent. Corolla pink or mauve, glabrous, up to 9 mm. long ; lobes 2–3 mm. long, obtuse or emarginate. Filaments about 0·4 mm. long ; anthers about 1·5 mm. long. Capsule globose, about 2 mm. in diameter.

TANGANYIKA. Mbeya District : Kikondo, 22 Oct. 1956, *Richards* 6714! ; Rungwe District : Rungwe Forest, 16 Sept. 1932, *Geilinger* 2365 !
DISTR. **T**7 ; not known elsewhere.
HAB. Rocky, shady places ; 1000–2250 m.

2. **A. sibthorpioïdes** *Hook. f.* in J.L.S. 7 : 205, t. 1 (1864) ; F.T.A. 3 : 488 (1877) ; Pax & Knuth in E.P. IV, 237 : 224, fig. 50 (1905) ; De Wild., Pl. Bequaert. 2 : 92 (1923) ; F.W.T.A. 2 : 184 (1931) ; Weim. in Svensk Bot. Tidskr. 30 : 39 (1936) ; P. Tayl. in K.B. 1958 : 147 (1958). Type : Cameroon Mt., *Mann* 2022 (K, holo.!)

Stems slender, prostrate, up to 80 cm. long, rooting at the nodes ; internodes 5–30 mm. long. Leaves numerous ; petiole slender, pubescent, 5–40 mm. long ; lamina broadly ovate or orbicular, up to 40 × 40 mm., sparsely pubescent on both surfaces, subacute, basally cordate, 5–7-crenate, the crenations coarsely dentate and the teeth short, broad, acuminate. Racemes 1–3(5)-flowered ; axes very short ; peduncles very short or obsolete, arising at intervals along the stem near the leaf-bases ; pedicels filiform, 5–8 mm. long ; bracts linear-lanceolate, about 3 mm. long. Calyx-segments lanceolate to broadly ovate-deltoid, basally rounded, hastate or cordate, acute, ciliate, almost as long as the corolla at anthesis, accrescent. Corolla white, 3–5 mm. long ; lobes 1–2 mm. long, rounded at apex, ciliate. Filaments about 0·5 mm. long ; anthers about 1·2 mm. long, rounded or cordate at the base. Capsule globose, sparsely pubescent, about 3 mm. in diameter, dehiscing by 5 thickened cartilaginous valves. Seeds 3-angled, minutely papillose, 0·8–1·0 mm. long. Fig. 1/12, 13.

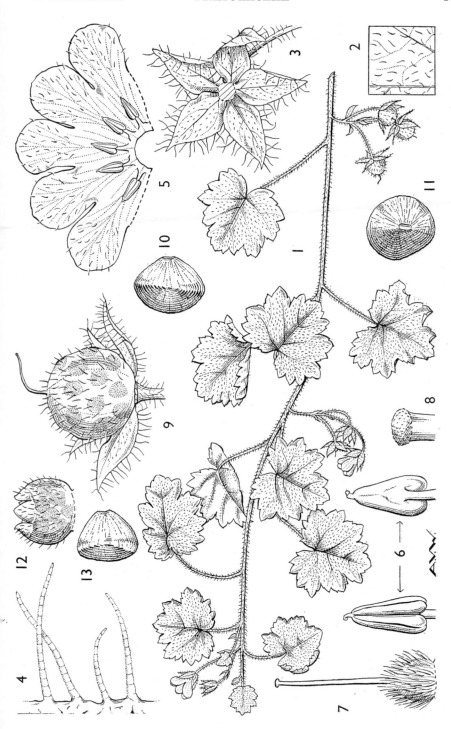

FIG. 1. *ARDISIANDRA WETTSTEINII*, from *Drummond & Hemsley* 4284—**1**, part of plant, × 1 ; **2**, surface of leaf, × 4 ; **3**, calyx, × 5 ; **4**, cilia on margin of calyx-lobes, × 30 ; **5**, corolla, opened out, × 5 ; **6**, anther, back and front view, × 15 ; **7**, pistil, × 10 ; **8**, stigma, × 40 ; **9**, fruit, × 5 ; **10**, seed, lateral view, × 20 ; **11**, seed, basal view, × 20. *A. SIBTHORPIOIDES*, from *Purseglove* 3746—**12**, fruit, × 5 ; **13**, seed, lateral view, × 20.

Uganda. Ruwenzori, Bwamba Pass, July 1940, *Eggeling* 4003 ! ; Kigezi District :
  Kachwekano Farm, Dec. 1951, *Purseglove* 3746 !
Kenya. Aberdare Mts. : Kinangop, Apr. 1938, *Chandler* 2227 !
Tanganyika. Lushoto District : Amani, 31 Dec. 1928, *Greenway* 1076 ! ; Morogoro
  District : Towero, Sept. 1930, *Haarer* 1896 ! ; Ulanga District : Mahenge, 23 Mar.
  1932, *Schlieben* 1953 !
Distr. U2 ; K3 ; T3, 6 ; Cameroon Mt. and Fernando Po, Belgian Congo, Ethiopia
  and the Sudan (Imatong Mts.).
Hab. Lowland rain-forest, upland dry evergreen forest and upland evergreen bushland,
  in rocky places and by streams ; 900–2670 m.

Syn. *A. engleri* Weim. in Svensk Bot. Tidskr. 30 : 40 (1936) ; Staner in Ann. Soc.
    Scient. Brux., sér. B, 56 : 244 (1936) ; F.P.N.A. 2 : 37 (1947). Type :
    Tanganyika, Lushoto District, Amani, *Engler* 711 (B, holo.†)
  A. engleri Weim. var. *microphylla* Weim. in Svensk Bot. Tidskr. 30 : 41 (1936).
    Type : Tanganyika, Morogoro District, Uluguru Mts., Feb. 1933, *Schlieben*
    3441 (BM, iso. !)

3. **A. wettsteinii** *R. Wagner* in Anzeiger Akad. Wiss. Wien 69 : 185 (1932) ;
P. Tayl. in K.B. 1958 : 147 (1958). Type : Tanganyika, Kilimanjaro,
Greenway 3862 (K, holoneo. !, EA, isoneo. !)

Stems slender, reddish, prostrate, up to 45 cm. long, rooting at the nodes ;
internodes 5–30 mm. long. Leaves numerous ; petiole slender, pubescent,
up to 4 cm. long ; lamina orbicular, up to 40 × 40 mm., sparsely pubescent on
both surfaces, obtuse, basally cordate, 7–9-crenate, the crenations with a
few short broad acute teeth. Racemes 1–6-flowered ; axes up to 10 mm. ;
peduncles up to 10 mm. ; pedicels filiform, up to 20 mm. long ; bracts
linear-lanceolate, 3–4 mm. long. Calyx-segments narrowly ovate, pubescent,
basally rounded, acute, ciliate, slightly accrescent, at anthesis about half as
long as the corolla. Corolla white or pale mauve, up to 8 mm. long ; lobes
2–3 mm. long, glabrous ; apex rounded or emarginate. Filaments about
0·6 mm. long ; anthers about 1·3 mm. long. Capsule globose, about 4–5 mm.
in diameter, indehiscent, with a thin transparent pericarp through which
can be seen the swollen whitish placenta and black seeds. Seeds about
1·2 mm. long. Fig. 1/1–11, p. 3.

Uganda. Rumenzori, Nyinabitaba, Aug. 1933, *Eggeling* 1390 ! ; Kigezi District :
  Muhavura-Mgahinga saddle, Sept. 1946, *Purseglove* 2147 !
Kenya. Mt. Kenya : West Kenya Forest Station, 3 Jan. 1922, *Fries* 714 ! ; Teita
  District : Teita Hills, Yale Peak, 13 Sept. 1953, *Drummond & Hemsley* 4284 !
Tanganyika. Kilimanjaro, Bismarck Hill, Feb. 1934, *Greenway* 3862 ! ; Lushoto
  District : Usambara Mts., Bumbuli Mission, 10 May, 1953, *Drummond & Hemsley*
  2486 ! Rungwe District : Mt. Rungwe, May 1953, *Eggeling* 6527 !
Distr. U2 ; K3, 4, 7 ; T2, 3, 7, 8 ; mountains of eastern Africa from Uganda and the
  adjacent parts of the Belgian Congo to Southern Rhodesia (Inyanga).
Hab. Upland rain-forest, upland dry evergreen forest, moist bamboo thickets and
  upland grasslands in shady damp places ; 1800–3600 m.

Syn. *A. orientalis* Weim. in Svensk Bot. Tidskr. 30 : 41 (1936) ; F.P.N.A. 2 : 38
    (1947). Types : Kenya, Mt. Kenya, *Fries* 714 & Aberdare Mts., *Fries*
    2422 ; Tanganyika, Kilimanjaro, *Haarer* 1002 & 1112 (all K, syn. !)
  A. orientalis Weim. var. *hirsuta* Weim. in Svensk Bot. Tidskr. 30 : 44 (1936) ;
    F.P.N.A. 2 : 38 (1937). Types : Belgian Congo, Mt. Nyiragongo, *Mildbraed*
    1389 & Mt. Muhavura, *Mildbraed* 1840 (both B, syn. †)
  A. stolzii Weim. in Svensk Bot. Tidskr. 30 : 44 (1936). Type : Tanganyika,
    Rungwe District, Kyimbila, *Stolz* 167 (BM, UPS, iso. !)

Note. This species is very similar indeed to *A. sibthorpioïdes* and in the absence of
fruits certain identification is almost impossible. The primary lobes of the leaf-
margin in the present species seem to be more (7–9) than in *A. sibthorpioïdes* (5–7),
but it is often difficult to distinguish between the primary lobes and the secondary
toothing and determine the number of the former.

2. LYSIMACHIA
L., Sp. Pl. : 146 (1753) & Gen. Pl. ed. 5 : 72 (1754)

Perennial herbs or rarely shrubs, prostrate or erect. Leaves alternate, opposite or verticillate, entire. Flowers solitary, axillary, racemose or paniculate. Calyx more or less deeply 5–6-lobed. Corolla rotate or subcampanulate, 5–6-lobed, white, yellow or rarely (as in the tropical African species) pink or purple. Stamens inserted on the corolla-tube ; staminodes often present and alternating with them. Ovary globose or ovoid ; style filiform. Capsule dehiscing by 5 valves, few- to many-seeded. Seeds various, usually angular.

A very large and widespread genus. The tropical African species belong to a small group of closely related species, the other members comprising the group occurring in the Mediterranean Region and eastern Asia.

Pedicels less than 5 mm. long ; style about 1 mm. long
(never more than 1·5 mm.) ; leaves very variable in
shape and size, usually more or less elliptic, petiolate
or more rarely sessile, surface provided with numerous
more or less conspicuous punctate glands .   .   1. *L. ruhmeriana*
Pedicels 10–20 mm. long ; styles more than 2 mm. long :
Stamens shorter than the corolla-lobes ; leaves lanceolate, sessile, more or less auriculate at the base,
surface without punctate glands   .   .   .   2. *L. volkensii*
Stamens longer than the corolla-lobes ; leaves narrowly elliptic .   .   .   .   3. *L. sp. A*

1. **L. ruhmeriana** *Vatke* in Linnaea, 40 : 204 (1876) : F.T.A. 3 : 489 (1877) ; Pax & Knuth in E.P. IV. 237 : 292 (1905) ; P. Tayl. in K.B. 1958 : 142 (1958). Type : Ethiopia, without locality, *Schimper* 1231 (B, holo. †, K, iso. !)

Rootstock short and woody, with very numerous wiry roots. Stems erect or decumbent, few to very many, up to 1 m. high, terete, usually strongly tinged with red below. Leaves alternate or subopposite, varying from broadly elliptic to lanceolate, acute, basally cuneate, obtuse or auriculate, green above, glaucous beneath, with numerous irregular-shaped punctate blackish glands. Flowers in congested or lax terminal racemes from 1 to 50 cm. in length ; pedicels 1–3 (–5) cm. long. Calyx-lobes oblong, 2–3 mm. long at anthesis, 3–4 mm. long in fruit ; apex obtuse, acute or rarely acuminate. Corolla subcampanulate, white or pink, about equal to or slightly longer than the calyx. Capsule globose, 3–4 mm. in diameter ; style persistent. Seeds numerous, 3-angled, about 1 mm. long. Fig. 2, p. 6.

Uganda. Kigezi District : Kachwekano Farm, Dec. 1949, *Purseglove* 3143 ! ; Busoga District : about 13 km. W. of Kamuli, 27 May 1953, *G. H. S. Wood* 764 ! ; Entebbe, Nov. 1922, *Maitland* 561 !
Kenya. Aberdare Mts. : Kinangop, 20 Dec. 1930, *Napier* 612 ! ; Mt. Kenya, 27 Dec. 1921, *Fries* 403 ! ; Elgon, 22 Feb. 1935, *G. Taylor* 3559 !
Tanganyika. Kilimanjaro, 27 Feb. 1934, *Greenway* 3832 ! ; Njombe District : Ndumbi Forest, Feb. 1954, *Paulo* 254 ! ; Songea District : Matengo Hills, 1 Mar. 1956, *Milne-Redhead & Taylor* 8780 !
Distr. U2–4 ; K1, 3–6 ; T1–3, 7, 8 ; from Ethiopia southwards to Natal and Madagascar, also in the Cameroons.
Hab. Riversides, marshes, ditches and damp places generally ; 960–3450 m.

Syn. *L. parviflora* Baker in J.L.S. 20 : 196 (1883) ; Pax & Knuth in E.P. IV. 237 : 291 (1905). Types : Madagascar, *Baron* 2303 & 3320 (both K, syn. !)
   L. africana Engl., P.O.A. C : 304 (1895) ; Pax & Knuth in E.P. IV. 237 : 291 (1905) ; Knuth & Mildbr. in Z.A.E. 518 (1913) ; F.P.N.A. 2 : 40 (1947) ; Andr., Fl. Pl. Sudan 3 : 66 (1956). Type : Tanganyika, Usambara Mts., *Holst* 9013A (B, holo. †, K, iso. !)

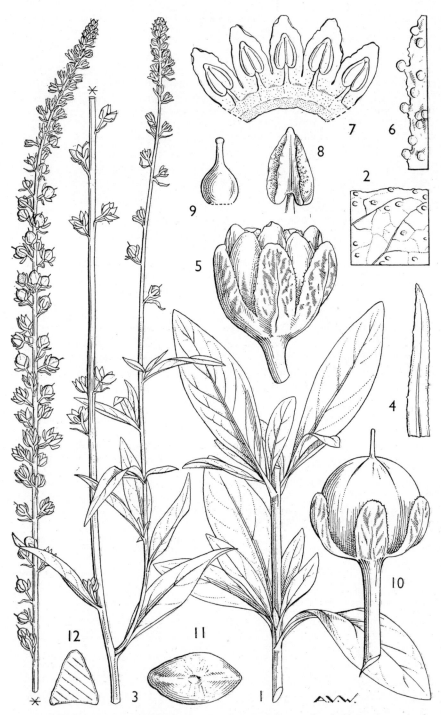

Fig. 2. *LYSIMACHIA RUHMERIANA*, from *Conrads* 5407—**1**, leaves and part of lower stem, × 1; **2**, surface of leaf, × 5; **3**, part of inflorescence, × 1; **4**, bract, × 8; **5**, flower, × 8; **6**, margin of calyx-lobe, × 80; **7**, corolla, opened out, × 10; **8**, anther, × 20; **9**, pistil, × 10; **10**, fruit, × 8; **11**, seed, basal view, × 30; **12**, vertical section (inverted) across the seed through the hilum, × 30.

L. *saganeitensis* Pax & Knuth in E.P. IV. 237 : 292 (1905). Type : Eritrea, Saganeiti, *Schweinfurth & Riva* 1684 (B, holo. †, K, iso. !)

L. *woodii* Pax & Knuth in E.P. IV. 237 : 292 (1905). Type : Natal, Van Reenens Pass, *Medley Wood* 4522 (B, holo. †, K, iso. !)

VARIATION. This species is very variable, especially in the length of inflorescence and shape and size of leaf. The species reduced above are mostly based on extreme forms but numerous intermediates make it impossible to distinguish them as distinct taxa in any rank.

2. L. volkensii *Engl.*, P.O.A. C : 304 (1895) ; Pax & Knuth in E.P. IV. 237 : 291 (1905) ; P. Tayl. in K.B. 1958 : 144 (1958). Type : Tanganyika, Kilimanjaro, Marangu, *Volkens* 533 (B, holo. †, K, iso. !)

Stems few to numerous, erect or decumbent, up to 1 m. high. Leaves subopposite or alternate, sessile, narrowly lanceolate, acute, auriculate at the base. Raceme 2–25 cm. long, usually congested at anthesis, lax in fruit ; pedicels ascending, up to 5 mm. long at anthesis, elongating in fruit up to 20 mm. Calyx-lobes oblong, obtuse or acute, 3–4 mm. long at anthesis, 4–5 mm. long in fruit. Corolla as long as or longer than the calyx. Style 2–3 mm. long. Capsule globuse, 3–4 mm. in diameter.

KENYA. N. Nyeri District : Nanyuki, 19 June 1943, *Moreau* 32 ! ; Kericho District , Sotik, 13 June 1950, *Bally* 7833 ! ; Masai District : Ngong Hills, 28 July 1956: *Milne-Redhead & Taylor* 11308 !

TANGANYIKA. Mbulu District : Mbulumbul, 26 June 1945, *Greenway* 7472 !

DISTR. **K**3–6 ; **T**2 ; not known elsewhere.

HAB. Upland grassland ; 1200–2400 m.

Imperfectly known species

3. L. sp. A

Probably annual ; plant seen damaged at base and branched ; branches ascending, rather stout ; leaves sessile, narrowly elliptic, up to 8 × 2 cm. Racemes 4–20 cm. long ; pedicels ascending, about 5 cm. long at anthesis, elongating up to 15 cm. long in fruit. Calyx-lobes oblong, about 2 mm. at anthesis and 3 mm. long in fruit. Corolla about 2·5 mm. long. Filaments 3 mm. long ; style about 2 mm. long. Capsule 3–4 mm. in diameter.

KENYA. Northern Frontier Province : Marsabit or Mt. Kulal [notes lost, exact locality not known] July 1934, *W. H. R. Martin* 209 !

DISTR. **K**1 ; known only from the above-cited gathering.

NOTE. Whilst this specimen (a single plant) may represent an undescribed species it is felt that it may be a very abnormal form of *L. ruhmeriana*. In its exserted anthers it resembles *L. atropurpurea* L. (an eastern Mediterranean species) but in other ways (e.g. corolla and style-length) it is quite different.

3. **ASTEROLINON**

Hoffmg. & Link, Fl. Portug. 1 : 332 (1820)

Glabrous annual herbs ; stems erect, simple or usually much branched. Leaves sessile, opposite, lanceolate or ovate. Flowers solitary, axillary, pedicellate, erect at anthesis, strongly recurved in fruit. Calxy deeply 5-lobed ; lobes lanceolate, acute. Corolla rotate-campanulate, 5-lobed, minute ; lobes orbicular or elliptic, much shorter than the calyx. Stamens inserted at the base of the corolla, shorter than the corolla-lobes ; filaments filiform ; anthers globose. Ovary globose ; style filiform. Capsule globose, dehiscing by 5 valves, 2–many-seeded. Seeds 3-angled.

A genus of two species, closely related to *Lysimachia*. The other species occurs throughout the Mediterranean region.

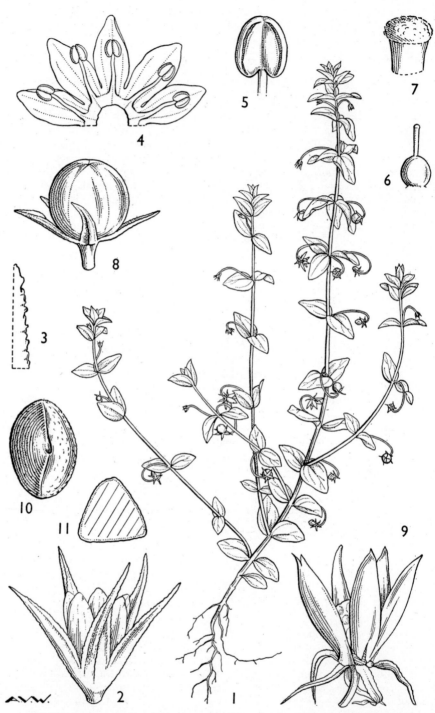

FIG. 3. *ASTEROLINON ADOËNSE*, from *Bogdan* 3181—**1,** plant, × 1 ; **2,** flower, × 12 ; **3,** margin of calyx-lobe, × 40 ; **4,** corolla, opened out, × 12 ; **5,** anther, × 48 ; **6,** pistil, × 12 ; **7,** stigma, × 72 ; **8,** fruit, × 8 ; **9,** fruit after dehiscence, × 8 ; **10,** seed, basal view, × 24 ; **11,** vertical section (inverted) across the seed through the hilum, × 24.

A. adoënse *Kunze* in Linnaea, 20 : 37 (1847) ; Pax & Knuth in E.P. IV. 237 : 317 (1905). Type : Ethiopia, Tigré, Aduwa, *Schimper* 63 (B, holo. †. K, iso. !)

Stems, except in very depauperate plants, much branched from the base, up to 30 cm. high, 4-angled. Leaves ovate, up to 12 × 8 mm. acute, basally rounded or subcordate. Pedicels capillary, up to 20 mm. long. Calyx-lobes lanceolate, acute, about 3 mm. long, with a narrow hyaline margin. Corolla-lobes elliptic, 1·5–2·0 mm. long and 1 mm. wide, rounded ; obscurely 5-nerved. Stamens about 1 mm. long. Ovary about 0·5 mm. long ; style about 0·8 mm. long. Capsule about 3 mm. in diameter. Seeds 1 mm. long, minutely papillose. Fig. 3.

UGANDA. Mbale District : Kapchorwa, 9 Sept. 1954, *Lind* 293 !
KENYA. Elgon, Sept. 1950, *Tweedie* 855 ! ; Nakuru District : near Molo, 24 July 1951, *Bogdan* 3181 ! ; Mt. Kenya, 16 Jan. 1922, *Fries* 959 !
TANGANYIKA. Kilimanjaro : Marangu, Nov. 1893, *Volkens* 1719 ! ; Mashame, Aug. 1927, *Haarer* 360 !
DISTR. **U**3 ; **K**3–5 ; **T**2 ; Ethiopia and the Sudan.
HAB. Upland grassland and as a weed of cultivation ; 1500–2700 m.

SYN. *Lysimachia adoënsis* (Kunze) Klatt in Abh. Naturw. Ver. Hamb. 4 (4) : 38 (1866) ; F.T.A. 3 : 489 (1877).
    [*Anagallis arvensis* sensu Duby in DC., Prodr. 8 : 69 (1844) & A. Rich., Tent. Fl. Abyss. 2 : 16 (1851) quoad spec. *Schimper* 63, *non* L.]

4. ANAGALLIS

L., Sp. Pl. : 148 (1753) & Gen. Pl., ed. 5 : 73 (1754) ; P. Tayl. in K.B. 1955 : 321 (1956) & 1958 : 133 (1958)

Glabrous annual or perennial herbs, erect or prostrate and rooting at the nodes. Leaves sessile or subpetiolate, alternate or opposite, capillary to orbicular, entire. Flowers small, pedicellate or rarely sessile, solitary and axillary or in terminal or subterminal racemes. Calyx-segments 5 (rarely 4 or 6), lanceolate ; margins aften hyaline. Corolla hypocrateriform to sub-campanulate ; lobes 5 (rarely 4 or 6), ½ to almost the length of the corolla, rounded or acuminate, sometimes glandular-ciliate. Filaments ± connate at the base (sometimes very shortly) all, or the lower part of, the connate part adnate to the corolla-tube, the free part often bearded with multi-cellular hairs ; anthers ovoid or oblong, obtuse or acute. Ovary globose, glabrous or covered with granular glands. Capsule globose or obovoid, indehiscent or circumscissile. Seeds 3-angled or globose, smooth or minutely papillose, 1–many.

A widespread genus, with, however, the majority of its species in North and tropical Africa.

Corolla-lobes about as broad as long, fringed with
    glands ; leaves opposite .   .   .   .   1. *A. arvensis* subsp.
                                                            arvensis

Corolla-lobes less than half as broad as long, not
    fringed with glands ; leaves opposite or alter-
    nate :
  Capsule obovoid, indehiscent, one-seeded ;
      corolla more than four times as long as calyx   2. *A. kingaënsis*
  Capsule globose, circumscissile, many-seeded ;
      corolla less than twice as long as calyx :
    Plant erect or ascending, not rooting at the
        nodes :

Leaves linear to filiform, more than ten
times as long as broad :
Corolla-lobes about 6 mm. long ; filaments
connate in their lower half above the
point of insertion on the corolla   .   3. *A. schliebenii*
Corolla-lobes under 4 mm. long ; filaments
not connate above the point of insertion on the corolla :
Filaments glabrous or with a few
minute papillae at the base ; corolla
pink   .   .   .   .   9. *A. rhodesiaca*
Filaments bearded ; corolla white   .   8. *A. acuminata*
Leaves lanceolate to ovate or obovate, less
than five times as long as broad :
Perennial with weakly erect stems up to
40 cm. long ; corolla-lobes more than
4 mm. long ; anthers more than
0·4 mm. long.   .   .   .   .   10. *A. tenuicaulis*
Annual, erect to 25 cm. ; corolla-lobes less
than 4 mm. long ; anthers less than
0·25 mm. long :
Leaves sessile, lanceolate to ovate ;
corolla equal to or exceeding the
calyx ; filaments filiform ; seeds
0·3–0·45 mm. long   .   .   11. *A. pumila*
Leaves petiolate, obovate-spathulate ;
corolla shorter than the calyx ;
filaments dilated ; seeds 0·45–0·6
mm. long   .   .   .   .   12. *A. djalonis*
Plant prostrate, rooting at the nodes or at least
the lower ones :
Leaves ovate, broadest below the middle ;
pedicels slender, usually exceeding the
leaves ; plant rooting at the lower
nodes only :
Corolla more than 9 mm. long, usually
hexamerous ; filaments bearded   .   7. *A. hexamera*
Corolla less than 6 mm. long, pentamerous ;
filaments glabrous .   .   .   .   6. *A. angustiloba*
Leaves oblanceolate, obovate or suborbicular, usually broadest at or above the
middle ; pedicels rather fleshy, usually
shorter than the leaves ; plant rooting
throughout the length of the stems :
Filaments connate in their lower third   .   4. *A. brevipes*
Filaments not connate .   .   .   .   5. *A. serpens* subsp.
                                      meyeri-johannis

1. **A. arvensis** *L.*, Sp. Pl. : 148 (1753) ; F.T.A. 3 : 490 (1877) ; Pax &
Knuth in E.P. IV. 237 : 323 (1905) ; P. Tayl. in K.B. 1955 : 329 (1956) &
1958 : 134 (1958). Type : presumably from Europe (LINN, syn. !)

Annual ; stems branching near the base and ascending or sometimes
simple and erect, quadrangular, strongly 4-winged, 5–50 cm. long. Leaves
opposite (or rarely verticillate), sessile, ovate, acute, subcordate at the base,
5–20 mm. long. Flowers solitary, axillary, pedicellate ; pedicels slender,
erect at anthesis becoming longer and recurved in fruit. Calyx-segments 5,

acute, about equal in length to the corolla-lobes. Corolla rotate, up to 15 mm. in diameter ; lobes up to as broad as long, apically entire, crenulate or denticulate, fringed with minute stalked glands. Filaments bearded ; anthers acute. Capsules about 5 mm. in diameter.

subsp. **arvensis**

Fruiting pedicels exceeding the leaves. Flowers red, pink or (usually, in tropical Africa) bright blue ; corolla-lobes broadly obovate, apically densely fringed with 3-celled glands.

KENYA. Elgon, July 1950, *Tweedie* 849 ! ; Kiambu District : Muguga, 28 Aug. 1952, *Verdcourt* 715 ! ; Machakos District : Kiu, 2 Mar. 1930, *Napier* 38 !
TANGANYIKA. Masai District : Ngorongoro, Apr. 1941, *Bally* 2323 ! Monduli, Great Ardai Plain, July 1943, *Greenway* 6757 ! ; Mbulu District : without locality, July 1943, *Moreau* 9082 !
DISTR. **K**1, 3, 4, 6 ; **T**2 ; an introduced weed, native in the Mediterranean Region and western Europe but now very widespread in the temperate regions of both hemispheres and, at higher altitudes, in the tropics.
HAB. Cultivated and waste land and dry upland evergreen bushland ; 1350–2450 m.

SYN. [*A. arvensis* L. var. *caerulea* sensu Pax & Knuth in E.P. IV. 237 : 323 (1905) partim, *non* (Schreb.) Gren. & Godr.]

NOTE. *A. arvensis* subsp. *foemina* (Mill.) Schinz & Thell. with shorter fruiting pedicels, narrower corolla-lobes which are invariably blue and have few 4-celled fringing glands, is not known from the area of this Flora.

2. **A. kingaënsis** *Engl.* in E.J. 30 : 371 (1902) ; P. Tayl. in K.B. 1955 : 348 (1956) & 1958 : 137 (1958). Type : Tanganyika, Njombe District, Kinga Mts., Kipengere, *Goetze* 958 (B, holo. †, BR, iso. !, K. photo-iso. !)

Perennial ; stems prostrate, simple or branched, crimson and rather fleshy, copiously rooting at the nodes, up to 40 cm. long. Leaves rather fleshy, alternate, broadly obovate or suborbicular, petiolate, rounded, ± cuneate at the base ; petiole up to 3 mm. long ; lamina up to 10 × 8 mm. Flowers few, axillary, solitary ; pedicels erect and about 1 mm. long in anthesis, reflexed and up to 20 mm. long in fruit. Calyx-segments ovate, acuminate, about 1 mm. long. Corolla pink, subcampanulate ; tube less than 1 mm. long ; lobes 4–6 mm. long and up to 1·5 mm. broad. Filaments filiform above, more or less connate and bearded in their lower half ; anthers obtuse. Fruit indehiscent, 1-seeded, obovoid-globose, about 1·6 mm. in diameter. Seed laterally compressed-globose, smooth, 1·5 mm. in diameter. Fig. 4/1–4, p. 12.

TANGANYIKA. Rungwe District : Upper Fishing Camp, Kiwira River, Aug. 1949, *Greenway* 8420 ! ; Njombe District : Elton Plateau, Ndumbi River, 12 Aug. 1956, *Richards* 8420 (fl.) ! & 14 May 1957, 9674 (fr.) !
DISTR. **T**7 ; not known elsewhere.
HAB. Streamsides in upland grassland ; 2100–2400 m.

3. **A. schliebenii** *Knuth & Mildbr.* in N.B.G.B. 11 : 672 (1932) ; P. Tayl. in K.B. 1955 : 331 (1956). Type : Tanganyika, Njombe District, Lupembe, *Schlieben* 1420 (BM, BR, EA, K, iso. !)

Perennial ; stems unbranched, erect, slender, angular. Leaves alternate, very numerous ; in the lower part of the stem filiform, up to 20 × 0·5 mm., apically gland-tipped ; in the upper part about 10 × 1·0 mm., acute. Flowers 3–10, axillary in the upper (but not the uppermost) part of the stem, forming an apparent short raceme or corymb. Pedicels capillary, up to 23 mm. long. Calyx-segments 5, narrowly lanceolate, acuminate, about 5 mm. long. Corolla 5-lobed almost to the base ; lobes oblong-ovate, rounded, about 7 mm. long, 3 mm. wide. Filaments about 5 mm. long, connate in their lower 2/5 in a tube, pubescent on the outer surface ; anthers

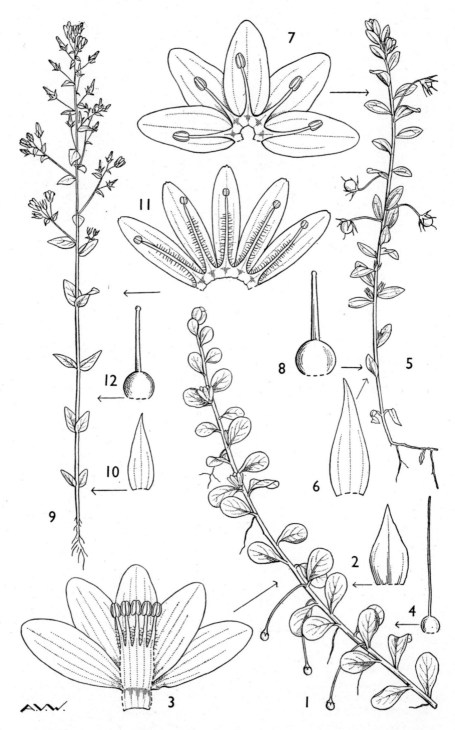

FIG. 4. *ANAGALLIS KINGAËNSIS*, from *Richards* 9674—**1,** part of plant, × 1 ; **2,** calyx-lobe, × 20 ; **3,** corolla, opened out, with attached stamens, × 6 ; **4,** pistil, × 8. *A. ANGUSTILOBA*, from *Drummond & Hemsley* 1569—**5,** part of plant, × 1 ; **6,** calyx-lobe, × 10 ; **7,** corolla, opened out, with attached stamens, × 6 ; **8,** pistil, × 10. *A. PUMILA* var. *BARBATA*, from *Milne-Redhead & Taylor* 9894— **9,** plant, × 1 ; **10,** calyx-lobe, × 8 ; **11,** corolla, opened out, with attached stamens, × 8 ; **12,** pistil, × 8.

about 0·6 mm. long, obtuse. Ovary globose, about 0·6 mm. in diameter ; style filiform, about 6 mm. long. Capsule not known.

TANGANYIKA. Njombe District : Lupembe, 12 Nov. 1931, *Schlieben* 1420 !
DISTR. **T**7 ; not known elsewhere.
HAB. Streambanks and marshes

4. **A. brevipes** *P. Tayl.* in K.B. 1958 : 135 (1958). Type : Tanganyika, Rungwe District, *Greenway & Brenan* 8278 (K, holo. !, EA, iso. !)

Perennial ; stems prostrate, much branched, rooting at lower nodes, up to 15 cm. long. Leaves fleshy, alternate, broadly obovate, rounded and shortly apiculate, cuneate at the base ; petiole 1–2 mm. long ; lamina up to 7 × 5 mm. Flowers pentamerous, rather numerous, axillary mainly in the upper parts of stems and branches. Pedicels 1–3 mm. long, erect at anthesis, reflexed in fruit. Calyx-segments narrowly lanceolate, keeled ; apex acute ; margin hyaline ; about as long as or slightly shorter than the corolla. Corolla subcampanulate, pale pink ; lobes narrowly oblong, 3–3·5 mm. long. Filaments glabrous, 2–2·2 mm. long, connate in their lower 2/5 in a tube, filiform above. Anthers obtuse, about 0·25 mm. long. Style about 1·8 mm. long. Fruit globose, about 2 mm. in diameter. Seeds 3-angled, 0·6–0·7 mm. long.

TANGANYIKA. Rungwe District : Upper Fishing Camp, Kiwira River, 25 Oct. 1947, *Greenway & Brenan* 8278 !
DISTR. **T**7 ; not known elsewhere.
HAB. Wet places in upland grassland ; 2250 m.

5. **A. serpens** [*Hochst. ex*] *DC.*, Prodr. 8 : 668 (1844) ; P. Tayl. in K.B. 1955 : 332 (1956). Type : Ethiopia, Silke, *Schimper* 547 (BM, K, iso. !)

Perennial (or ? annual) ; stems prostrate, closely adpressed to the ground and rooting at the nodes, simple or slightly branched, rather fleshy, pink, up to 40 cm. long. Leaves numerous, rather fleshy, spreading or suberect, opposite or alternate, usually ± obovate but varying from narrowly oblanceolate to suborbicular, acute or obtuse, cuneate into a short petiole ; lamina 4–17 mm. long. Flowers usually rather few, solitary, axillary, pedicellate ; pedicels erect, rather fleshy, often recurved in fruit, variable in length, up to 25 mm. Calyx-segments 5, lanceolate, acute, 2–7 mm. long. Corolla pink, campanulate, 3–15 mm. long, deeply 5-lobed ; lobes obovate, rounded. Filaments glabrous, filiform, sometimes somewhat dilated in the lower half, usually about ½ length of corolla ; anthers obtuse. Capsule globose, 2–5 mm. in diameter.

subsp. **meyeri-johannis** (*Engl.*) *P. Tayl.* in K.B. 1955 : 334 (1956). Type : Tanganyika, Kilimanjaro, *Meyer* 153 (B, holo. †)

Leaves alternate, very variable in size and shape but usually ± obovate.

UGANDA. Elgon, Apr. 1950, *Forbes* 275 ! & Dec. 1939, *A. S. Thomas* 2688 A !
KENYA. Aberdare Mts., June 1931, *Dent* 1304 ! ; Mau Range, Jan. 1946, *Bally* 4944 ! ; Mt. Kenya, Nov. 1943, *Bally* 3310 !
TANGANYIKA. Kilimanjaro, Feb. 1928, *Haarer* 1146 ! & Feb. 1934, *Greenway* 3741 !
DISTR. **U**3 ; **K**2–6 ; **T**2 ; known only from the higher mountains of East Africa.
HAB. In bogs and by streams on upland moor : 3000–4500 mm., descending, by streams, to 1920 m.

SYN. *A. quartiniana* (A. Rich.) Engl. var. *meyeri-johannis* Engl., Hochgeb. Trop. Afr. 330 (1892). Type : Tanganyika, Kilimanjaro, *Meyer* (B, holo. †)
A. meyeri-johannis (Engl.) Engl. in E.J. 30 : 372 (1902). Type : Tanganyika, Kilimanjaro, *Meyer* (B, holo. †)
A. kilimandscharica Knuth in E.P. IV. 237 : 326 (1905) ; Fries in N.B.G.B. 8 : 336 (1923). Type : Tanganyika, Kilimanjaro, *Volkens* 813 (EA, K, iso. !)
A. bella M.B. Scott in K.B. 1914 : 336 (1914) ; T.C.E. Fries in N.B.G.B. 8 : 335 (1923). Type : Kenya : Aberdare Mts., *Battiscombe* 833 (K, holo. !)

A. keniënsis T.C.E. Fries in N.B.G.B. 8 : 333 (1923). Type : Kenya, Mt. Kenya, *Fries* 1171 (UPS, holo. !, K, iso. !)

A. granvikii T.C.E. Fries in N.B.G.B. 8 : 335 (1923). Type : Kenya, Elgon, *Granvik* (S, holo. !)

A. roberti T.C.E. Fries in N.B.G.B. 8 : 337 (1923). Type : Kenya, Aberdare Mts., *Fries* 2261 (UPS, holo. !, K, iso. !)

A. iraruënsis T.C.E. Fries in N.B.G.B. 8 : 337 (1923). Type : Kenya, Mt. Kenya, *Fries* 1847 (UPS, holo. !, K, iso. !)

A. aberdarica T.C.E. Fries in N.B.G.B. 8 : 338 (1923). Type : Kenya, Aberdare Mts., *Fries* 2397 (UPS, holo. !, K, iso. !)

A. micrantha T.C.E. Fries in N.B.G.B. 8 : 338 (1923). Type : Kenya, Mt. Kenya, *Fries* 369 (UPS, holo. !, K, iso. !)

NOTE. *A. serpens* subsp. *serpens* does not occur in the area of this Flora but on the mountains to the north and south of it (in the Sudan and Ethiopia and in Southern Rhodesia respectively). It differs from subsp. *meyeri-johannis* in having opposite, always orbicular leaves.

VARIATION. This is an excessively variable species, mainly in the size of its flowers and the size and shape of its leaves. The very narrow-leaved forms (Aberdare Mts., *Battiscombe* 833, Marakwet Hills, *Dale* 3436) are quite strikingly different at first sight from both the very large forms (e.g. Mt. Kenya, *Fries* 1171) and from the very small ones (e.g. Aberdare Mts., *Fries* 2397). However all of these extremes are connected by intermediate forms.

6. **A. angustiloba** *(Engl.) Engl.* in E.J. 30 : 372 (1902) ; P. Tayl. in K.B. 1955 : 337 (1956).   Type : Tanganyika, Uluguru Mts., Lukwangule, *Drummond & Hemsley* 1569 (K, neo. !)

Perennial ; stems prostrate, simple or branched, rather fleshy, usually rooting at the lower nodes only, up to 50 cm. long. Leaves rather fleshy, opposite or alternate (often on the same plant) ovate, shortly petiolate, spreading, acute, rounded at the base, 8–15 × (2·5–) 4–8 mm. Flowers often numerous, axillary, solitary, pedicellate ; pedicels filiform, up to 20 mm. long, reflexed and elongated in fruit. Calyx-segments 5 (or rarely 6), lanceolate, acute, 3–4 mm. long. Corolla white or pale pink, campanulate, deeply 5- (rarely 6-) lobed ; lobes narrowly oblong, 4·5–6 (–8) mm. long, 5–7-nerved. Filaments filiform, glabrous, about $\frac{2}{3}$ as long as corolla ; anthers obtuse. Capsule globose, 2–3 mm. in diameter. Fig. 4/5–8, p. 12.

UGANDA. Ruwenzori, Bwamba Pass, Apr. 1942, *Eggeling* 4974 !; Kigezi District : Mt. Sabinio, June 1947, *Purseglove* 2441 ! ; Mt. Mgahinga, Oct. 1947, *Purseglove* 2521 !

KENYA. Mt. Kenya, Churi, Feb. 1922, *Fries* 1847a † ; Embu District : Ena River, 28 Sept. 1956, *Ossent* 171 !

TANGANYIKA. Lushoto District : W. Usambara Mts., Lukozi, May 1953, *Drummond & Hemsley* 2682 ! ; Morogoro District : Uluguru Mts., Lukwangule Plateau, Jan. 1934, *Michelmore* 910 & 915 ! !

DISTR. **U**2 ; **K**4 ; **T**3, 6 ; also in the Belgian Congo on the Virunga Mts.

HAB. Bogs, swamps, and by streamsides in upland grasslands, mainly 2100–2400 m. but descending by streamsides to 1500 m. and rarely ascending to 3000 m.

SYN. *A. quartiniana* (A. Rich.) Engl. var. *angustiloba* Engl. in E.J. 28 : 447 (1901). Type : Tanganyika, Uluguru Mts., Lukwangule, *Goetze* 294 (B, holo. †)

A. ruandensis Knuth & Mildbr. in Z.A.E. : 518 (1911) ; Fries in N.B.G.B. 8 : 332 (1923) ; F.P.N.A. 2 : 41 (1947). Type : Belgian Congo, Lake Kalago, *Mildbraed* 1556 (B, holo.·†)

A. churiënsis T.C.E. Fries in N.B.G.B. 8 : 332 (1923). Type : Mt. Kenya, *Fries* 1847a (UPS, holo.†)

A. kigesiënsis Good in J.B. 62 : 335 (1924). Type : Uganda, Behungi, *Godman* 214 (BM, holo. !)

A. ulugurensis Knuth in N.B.G.B. 12 : 88 (1934). Type : Tanganyika, Uluguru Mts., Lukwangule, *Schlieben* 3498 (BM, iso. !)

VARIATION. Specimens from the eastern part of the range tend to have alternate leaves whilst those from the west are most frequently opposite. However gatherings have been seen with alternate and opposite leaves from both places.

7. **A. hexamera** *P. Tayl.* in K.B. 1955 : 339 (1956) & 1958 : 137 (1958).
Type : Kenya, Ravine District, Timboroa, *Williams & Piers* 560 (K, holo. !,
EA, iso. !)

Perennial ; stems prostrate, simple or rarely branched, pink and rather
fleshy, rooting at the lower nodes, up to 60 cm. long. Leaves spreading,
rather fleshy, opposite or rarely alternate, ovate, shortly petiolate, acute,
rounded at the base ; petiole up to 2 mm. long ; lamina up to 25 × 10 mm.
Flowers rather few, axillary, solitary, long-pedicellate ; pedicels filiform,
erect at anthesis and up to 3 cm. long, reflexed and up to 6 cm. long in fruit.
Calyx-segments 6 (rarely 5 or up to 8), lanceolate, acute, 4·5–6 mm. long.
Corolla pink or white, subcampanulate, deeply 6- (rarely 5- or up to 8-)
lobed ; lobes oblong 9–12 mm. long, 3–4 mm. wide, 7–11-nerved, rounded.
Filaments about $\frac{3}{4}$ length of corolla, dilated and bearded in the lower $\frac{1}{3}$,
filiform above ; anthers obtuse. Capsule globose, about 3 mm. in diameter.
Seeds 0·8–1·0 mm. long.

KENYA. Nakuru District : Ol-joro-orok, Jan. 1932, *Pierce* 1656 ! ; Uasin Gishu
District : Uasin Gishu Plateau, *Dowson* 664 ! ; N. Nyeri District : Naro Moro, Aug.
1933, *Jex Blake* 6144 !
TANGANYIKA. Mbeya District : Chunya escarpment, Jan. 1957, *Richards* 7937
DISTR. **K**3–5 ; **T**7 ; also in Ethiopia.
HAB. Bogs and marshes in upland grassland ; 2100–2600 m.

8. **A. acuminata** [*Welw. ex*] *Schinz* in Bull. Herb. Boiss. 2 : 221 (1894) ;
P. Tayl. in K.B. 1955 : 341 (1956) & 1958 : 140 (1958). Type : Angola,
Huilla District, Morro de Lopollo, *Welwitsch* 275 (Z, holo., BM, K, iso. !)

Annual ; stems erect, simple or rarely branched above, 3–12 cm. high,
tinged reddish at base. Lower leaves opposite, upper alternate, narrowly
linear, 2–7 mm. long, less than 0·5 mm. wide, the upper ones acute, the lower
tipped with a small black gland. Flowers in a terminal corymbiform raceme
which elongates slightly in fruit ; pedicels ascending, filiform, up to 1 cm.
long. Calyx-segments narrowly lanceolate, 2–3 mm. long, acuminate and
with a narrow hyaline margin. Corolla white, about 3 mm. long ; lobes
(spreading in sunlight) narrowly oblong, about 0·6 mm. wide, apically
truncate and shortly mucronate or tridentate, 3-nerved. Filaments filiform,
about 2 mm. long, glabrous above, bearded below ; anthers globose, obtuse,
about 0·3 mm. long. Ovary sparsely papillose ; style about 1·2 mm. long.
Capsule about 1·5 mm. in diameter. Seeds numerous, about 0·35 mm. long.

TANGANYIKA. Songea District : near R. Mtanda about 9·5 km. SW. of Songea, 21 June
1956, *Milne-Redhead & Taylor* 10856 !
DISTR. **T**8 ; Angola, Northern and Southern Rhodesia.
HAB. Seasonally flooded places, flowering as the water-level subsides at the beginning
of the dry season ; 1000 m.

9. **A. rhodesiaca** *R. E. Fries*, Wiss. Ergebn. Schwed. Rhod. Congo Exped.
1911–12, 1 : 253 (1916) ; P. Tayl. in K.B. 1955 : 342 (1956) & 1958 : 140
(1958). Type : Northern Rhodesia, between Fort Rosebery and Lake
Bangweolo, *R. E. Fries* 634 (UPS, holo. !, K, iso. !)

Annual ; stems erect, simple or branched above, 6–20 cm. high, green or
± tinged reddish above, rather thick and fleshy below. Leaves alternate or
subopposite, narrowly linear, 2–8 mm. long, less than 0·5 mm. wide, the
upper ones acute, the lower tipped with a small black gland. Flowers in a
terminal raceme elongating in fruit ; pedicels ascending, filiform, up to 1 cm.
long. Calyx-segments lanceolate to lanceolate-ovate, 2–2·5 mm. long and
up to 1 mm. wide, acute and with narrow hyaline margin. Corolla pink,
about 3 mm. long ; lobes (opening but not spreading in sunlight) broadly

oblong, about 1 mm. wide, apically truncate, emarginate or rounded and sometimes very shortly mucronate or tridenticulate, 3–5-nerved. Filaments filiform, about 1 mm. long, glabrous or with a few minute papillae at the base ; anthers globose, obtuse, about 0·2 mm. long. Ovary densely covered with minute papillae ; style about 0·6 mm. long. Capsule about 1·5 mm. in diameter. Seeds numerous, about 0·3 mm. long.

TANGANYIKA. Songea District : Kwamponjore valley, about 9·5 km. SW. of Songea, 19 June 1956, *Milne-Redhead & Taylor* 10836 ! & near R. Mtanda about 9·5 km. SW. of Songea, 21 June 1956, *Milne-Redhead & Taylor* 10858 !
DISTR. **T8** ; also in Northern Rhodesia.
HAB. Seasonally flooded places, flowering as the water-level subsides at the beginning of the dry season ; 1000 m.

10. **A. tenuicaulis** *Baker* in J.B. 20 : 172 (1883) ; P. Tayl. in K.B. 1955 : 342 (1956) & 1958 : 140 (1958). Type : Madagascar, Betsileo, *Baron* 240 (K, holo. !)

Perennial ; stems decumbent, branched or rarely simple, distinctly winged, up to 40 cm. long. Lower leaves opposite or subopposite, upper alternate, sessile or subsessile, broadly ovate, acute or shortly acuminate, basally rounded or subcordate, 3–10 mm. long. Flowers in few- to many-flowered terminal racemes up to 15 cm. long ; pedicels spreading, filiform, up to 15 mm. long. Calyx-segments 5, lanceolate, acuminate, 2·5–3·5 mm. long, 5-nerved. Corolla subcampanulate, white, subpersistent in fruit ; lobes 5, narrowly oblong ; apex truncate, shortly mucronate to tridentate, 4–5 mm. long, 5-nerved. Filaments filiform, bearded in the lower half. Anthers oblong, obtuse, 0·4–0·7 mm. long, Capsule about 1·5 mm. in diameter. Seeds 0·5–0·6 mm. long.

UGANDA. West Nile District : Logiri, *Eggeling* 1876 ! ; Masaka District : Lake Nabugabo, Oct. 1932, *Eggeling* 962 !
KENYA. Uasin Gishu District : Kipkarren, Sept. 1931, *Brodhurst-Hill* 285 ! ; S. Kavirondo District : Lolgorien, Sept. 1935, *Napier* 6098 !
TANGANYIKA. Without locality, *Hannington* ! ; Songea District : 12 km. E. of Songea, Dec. 1955, *Milne-Redhead & Taylor* 7767 !
DISTR. **U**1, 3, 4 ; **K**3, 5 ; **T**?5, 8 ; southwards to Southern Rhodesia, Natal and Madagascar.
HAB. Bogs, swamps and damp grassland ; 1000–1800 m.

SYN. *A. hanningtonii* Baker in K.B. 1901 : 127 (1901). Types : Tanganyika, without locality, *Hannington* (K, syn.!) ; Northern Rhodesia, Abercorn District, Fwambo, *Carson* 45 & 72 (K, syn. !)
    A. pumila Sw., var. *natalensis* Schltr. ex Knuth in E.P. IV. 237 : 332 (1905). Type : Natal, Inanda, *Medley Wood* 1609 (BM, K, iso. !)

11. **A. pumila** *Sw.*, Prodr. Veg. Ind. Occ. 1 : 40 (1788) ; Hutch. & Dalz., F.W.T.A. 2 : 184 (1931), partim, excl. syn. *A. djalonis ;* P. Tayl. in K.B. 1955 : 342 (1956), partim, excl. var. *djalonis ;* & 1958 : 140 (1958). Type : Jamaica, *Swartz* (BM, iso. !)

Annual ; stem erect, simple or often branched above, angular, up to 25 cm. tall. Leaves alternate or subopposite, sessile, narrowly elliptic to broadly ovate, acute, basally cuneate, rounded or rarely subcordate, up to 10 mm. long. Flowers in terminal racemes occupying the upper half of the stem and branches. Pedicels filiform, ascending at anthesis, more or less spreading in fruit, up to 15 mm. long. Calyx-segments 5, rarely 4, lanceolate, acuminate, 1·5–3 mm. long. Corolla white, persistent and reddish in fruit, subcampanulate ; lobes 5, rarely 4, narrowly oblong, acute, acuminate, truncate-apiculate or tridentate, 3-nerved, 1·7–4 mm. long. Filaments shortly adnate to the corolla-tube at the base ; free part filiform, glabrous or bearded. Anthers ovoid, obtuse, 0·25 mm. long or less. Seeds 3-angled, 0·3–0·45 mm. long.

var. **pumila**

Corolla-lobes 1·7–3 mm. long, scarcely exceeding the calyx. Filaments glabrous.

KENYA. Nairobi, Thika Road House, July 1951, *Verdcourt* 534 (partly) !

TANGANYIKA. Iringa District : Iringa, 13 July 1956, *Milne-Redhead & Taylor* 11138 (partly) ! ; Songea District : Lumecha Bridge, 4 May 1956, *Milne-Redhead & Taylor* 9893 !

DISTR. **K**4 ; **T**7, 8 ; scattered throughout tropical Africa, India, Malaya, Australia and America (peninsular Florida southwards to Paraguay).

var. **barbata** *P. Tayl.* in K.B. 1955 : 345 (1956) & 1958 : 141 (1958). Type : Tanganyika, Tanga District, Magunga Estate, Aug. 1953, *Faulkner* 1230 (K, holo. !, UPS, iso. !)

Corolla-lobes 2·5–4 mm. long, usually considerably longer than the calyx. Filaments bearded. Fig. 4/9–12, p, 12.

UGANDA. Teso District : Omunyal Swamp, 14 Sept. 1954, *Lind* 358 ! Mbale District : Kapchorwa, 7 Sept. 1954, *Lind* 246 !

KENYA. Nairobi, *Dowson* 406 ! Kiambu, July 1932, *Mainwaring* 2164 !

TANGANYIKA. Mwanza District : Ukerewe Island, July 1928, *Conrads* 900 ! ; Arusha District : Moshi, *Haarer* 1530 ! ; Songea District : Lumecha Bridge, 4 May 1956, *Milne-Redhead & Taylor* 9894 !

ZANZIBAR. Zanzibar Is., Kidoti, *Hildebrandt* 1147 !

DISTR. **U**3 ; **K**4 ; **T**1–3, 7, 8 ; **Z** ; more or less throughout tropical Africa.

HAB. (of species as whole). Seasonally flooded ground and damp places generally ; sea level to 2100 m.

Intermediate forms between var. *pumila* and var. *barbata* occasionally occur.

12. **A. djalonis** *A. Chev.* in Journ. de Bot., sér. 2, 22 : 115 (1909) & in Expl. Bot. 1 : 384 (1920) ; P. Tayl. in K.B. 1958 : 141 (1958). Type : French Guinea, Fouta Djallon, *Chevalier* 18876 (P, holo. !, K, iso. !)

Annual ; stem erect, simple or branched from near the base, angular, with the angles narrowly winged, up to 20 cm. tall (in West Africa, usually 5–10 cm. in East Africa). Leaves alternate, obovate-spathulate, shortly petiolate ; petiole 2–5 mm. long ; lamina up to 15 × 10 mm. Flowers solitary, axillary, numerous throughout the length of the stem and branches. Pedicels filiform, ascending at anthesis, spreading in fruit, up to 20 mm. long. Calyx-segments 5 or 4, lanceolate, acuminate, about 2 mm. long. Corolla white, persistent and reddish in fruit ; lobes 5 or 4, oblong, rounded or sub-acute, sometimes very obscurely denticulate, 3-nerved, 1–1·5 mm. long. Filaments adnate to the corolla-tube for about $\frac{1}{3}$–$\frac{1}{2}$ their length ; free part dilated, deltoid, glabrous ; anthers obtuse. Seeds 3-angled, 0·45–0·6 mm. long.

KENYA. Uasin Gishu District : Kipkarren, Oct. 1931, *Brodhurst-Hill* 551 !

TANGANYIKA. Rungwe District : Rungwe, Sept. 1932, *Geilinger* 2467 ! Songea District : Lukumburu, 6 July 1956, *Milne-Redhead & Taylor* 10750 !

DISTR. **K**3 ; **T**7, 8 ; also in French Guinea, Northern Rhodesia and Angola.

HAB. Damp bare earth ; 1000–1800 m.

SYN. *Anagallis pumila* Sw. var. *djalonis* (A. Chev.) P. Tayl. in K.B. 1955 : 346 (1953).
[*A. pumila* sensu Hutch. & Dalz., F.W.T.A. 2 : 184 (1931), partim, quoad syn. et spec. *Chevalier* 20193, *non* Swartz]

5. SAMOLUS

L., Sp. Pl. : 171 (1753) & Gen. Pl. ed. 5 : 78 (1754)

Annual or perennial herbs with erect or rarely prostrate stems. Leaves alternate, cauline or forming a basal rosette, linear to spathulate, entire. Flowers in terminal racemes or corymbs ; bracts at the base or near the middle of the pedicels. Calyx-tube adnate to the ovary in its lower half ; limb 5-lobed, persistent. Corolla perigynous, subcampanulate ; limb 5-lobed. Stamens inserted on the corolla-tube ; subulate staminodes often

FIG. 5. *SAMOLUS VALERANDI*, from *Napier* 749—**1**, part of plant, × 1 ; **2**, flower, × 12 ; **3**, corolla, opened out, with attached stamens and staminodes, × 10 ; **4**, pistil, with two calyx-lobes removed, × 12 ; **5**, fruit, × 12 ; **6**, seed, × 40.

present alternating with the stamens. Ovary subglobose ; style short. Capsule globose, many-seeded, dehiscing by 5 valves. Seeds numerous, angular.

A small widespread genus found mainly in the southern hemisphere with the following cosmopolitan species occurring in practically every region of the world.

S. valerandi *L.*, Sp. Pl. : 171 (1753) ; F.T.A. 3 : 490 (1877) ; Pax & Knuth in E.P. IV. 237 : 336 (1905). Type : presumably from Europe, Hortus Cliffortianus (BM, syn. !)

Glabrous annual herb with erect stems to 50 cm. high, simple or branched above. Leaves in a more or less well-developed basal rosette and more or less numerous throughout the length of the stem, the lowermost spathulate and up to 12 cm. long, those above obovate or elliptic, shorter. Flowers in terminal racemes up to 20 cm. long ; pedicels ascending, up to 25 mm. long, more or less geniculate at the insertion of the bract at or slightly above the middle. Calyx about 2 mm. long ; lobes about half that length, ovate-deltoid, acute. Corolla about 2 mm. long ; lobes about $\frac{2}{3}$ of that length, obovate. Stamens inserted at the base of the corolla-tube, about 1 mm. long ; staminodes minute, about 0·2 mm. long. Capsule-valves strongly reflexed after dehiscence. Seeds about 0·6 mm. long. Fig. 5.

Kenya. Masai District : Kajiado, 9 Jan. 1931, *Napier* 749 ! Machakos District : Simba River, 21 Apr. 1902, *Kassner* 640 !
Distr. **K**4, 6 ; scattered throughout the whole of Africa and most of the rest of the world.
Hab. River-banks and drying up river-beds ; 1710 m.

INDEX TO PRIMULACEAE